REWARD

REWARD

Starter

Grammar and Vocabulary Workbook

Simon Greenall

MACMILLAN
HEINEMANN
English Language Teaching

Macmillan Heinemann English Language Teaching, Oxford

A division of Macmillan Publishers Limited

Companies and representatives throughout the world

ISBN 0 435 24268 7 (with key)

ISBN 0 333 74254 0 (without key)

Text © Simon Greenall 1998

Design and illustration © Macmillan Publishers Limited 1998

Heinemann is a registered trademark of Reed Educational and Professional Publishing Limited

First published 1998

Illustrated by Paul Collicutt

Printed and bound in Great Britain by Redwood Books, Trowbridge, Wiltshire

2003 2002 2001 2000 1999
11 10 9 8 7 6 5 4 3 2

Contents

Lessons 1-5

GRAMMAR

1 Complete the sentences with *am*, *are* or *is*.

Are you from Madrid?

1 _____ you a journalist? Yes, I _____ .

2 Jack _____ from Britain.

3 Olga _____ Russian.

4 I _____ a teacher.

5 What _____ your job?

6 You _____ a student from Thailand.

7 Harry _____ from Britain and Susan _____ from the USA.

8 _____ Kirk an actor?

9 She _____ an engineer.

2 Write the contracted forms.

He is from Washington.

He's from Washington.

1 You are Russian.

_____

2 She is a student.

_____

3 Steve is a doctor.

_____

4 I am from Turkey.

_____

5 Tina is an actress.

_____

6 What is your name?

_____

7 He is Brazilian.

_____

8 Monica is American and she is a singer.

_____

9 Paulo is a waiter and he is from Rome.

_____

10 What is your job?

_____

3 Put the conversations in the correct order.

1 a I'm very well, thanks. How are you? ☐

  b I'm fine, thank you. ☐

  c Hello, George. How are you? ☐

2 a I'm Pedro. What's your job, Maria? ☐

  b Oh, I'm a secretary. And what's your job, Pedro? ☐

  c Hello, what's your name? ☐

  d I'm a teacher. ☐

  e I'm Maria. What's your name? ☐

4 Punctuate.

whats your job

What's your job?

1 whats your name

_____

2 how are you im fine thanks

_____

3 johns from new york

_____

4 hes from tokyo and hes japanese

_____

5 youre an engineer

_____

6 Suzannes a teacher shes from france

_____

7 mr fords from london

_____

8 whats your telephone number

_____

5 Tick (✓) the correct sentence.

  a Are you neighbours? Yes, we're. ☐

  b Are you neighbours? Yes, we are. ☑

1 a Are you American? Yes, I'm. ☐

  b Are you American? Yes, I am. ☐

2 a Michael is a doctor. ☐

  b Michael is doctor. ☐

3 a Pamela is British. ☐

  b Pamela is Britain. ☐

4 a How you are? ☐

  b How are you? ☐

5 a What your telephone number? ☐

  b What's your telephone number? ☐

6 a Are you a student? No, I'm not. ☐

  b Are you a student? No, I amn't. ☐

6 Write *a* or *an*.

She's __*a*__ student.

1 She's _____ teacher.

2 Abdel is _____ actor.

3 Mrs Smith is _____ engineer.

4 You're _____ doctor.

5 I'm _____ actress.

6 He's _____ waiter.

7 Cindy's _____ journalist.

8 William's _____ singer.

7 Write the questions in the conversation.

A *Hello. How are you?* _____

B I'm fine, thanks.

A _____

B I'm Bruno.

A _____

B B-R-U-N-O.

A _____

B No, I'm not. I'm a teacher.

A _____

B My telephone number is 0189 45832.

8 Complete with *He's* or *She's*.

Bogdan is from Krakow. __*He's*__ Polish.

1 Jane's from Washington. _____ American.

2 Julia's a secretary. _____ from London.

3 Jorge is Brazilian. _____ a doctor.

4 Catherine is French. _____ from Lyon.

5 Charles is an actor. _____ from San Francisco.

6 Mr Cook is British. _____ from Edinburgh.

9 Put the conversations in the correct order.

1 a No, I'm not. I'm from Japan. Where are you from?. ☐

  b Are you from Thailand? ☐

  c I'm from Brazil. ☐

2 a Oh, how do you spell your name? ☐

  b Hello, are you Peter? ☐

  c L-A-C-H-L-A-N. ☐

  d No, I'm not. I'm Lachlan. ☐

10 Match.

1 How   a phone number?
2 She's from   b spell 'Reward'?
3 What's your c Turkish.
4 Thank you   d are you?
5 She's   e very much.
6 How do you   f Turkey.

11 Answer the questions for you. Use *Yes, I am* or *No, I'm not*.

1 Are you a student?

_____

2 Are you an actor?

_____

3 Are you from Rio?

_____

4 Are you French?

_____

5 Are you from Turkey?

_____

6 Are you a secretary?

_____

SOUNDS

12 Underline the stressed syllables.

What's your <u>job</u>? I'm an <u>ac</u>tor.

1 Are you a student?
2 I'm from Thailand.
3 What's your name?
4 What's your telephone number?
5 She's a doctor.
6 I'm very well, thanks.
7 Thank you very much.
8 He's from Japan.
9 I'm American and I'm an engineer.
10 She's from New York.

13 Put the words in the correct column according to their stress.

number underline punctuate
correct complete engineer listen journalist
waiter student Japanese Italy hello Brazil
telephone zero Japan

☐ ☐   ☐ ☐

number _____   _____

_____   _____

_____   _____

_____   _____

☐ ☐ ☐   ☐ ☐ ☐

_____   _____

_____   _____

_____   _____

_____   _____

14 Circle the different sound.

	C	P	Ⓨ	T
1	A	K	V	H
2	G	J	T	E
3	X	N	F	C
4	U	V	B	D
5	Z	L	M	R

VOCABULARY

15 Match the telephone numbers.

1 904987   a one nine three three, four five six two
2 464 3276   b one nine, three one two oh five five
3 464292   c nine oh four nine eight seven
4 1933 4562   d four six four, three two seven six
5 90 20 57   e nine oh, two oh, five seven
6 19 312055   f four six four two nine two

16 Look at the pictures and complete the 'job' puzzle. Find the hidden job.

		1						
	2							
3								
	4							
	5							

Hidden job _____

17 What are the countries?

Russia

1 _____

2 _____

3 _____

4 _____

5 _____

18 Circle the odd-one-out.

		Italian	Turkey	American
	British			
1	Russian	student	doctor	secretary
2	put in order	listen	very well	write
3	Thailand	Turkey	Brazilian	Japan
4	look	count	listen	job
5	nationality	waiter	engineer	singer
6	Japanese	Britain	Thai	American
7	Goodbye	Hello	Thanks	country
8	is	he	am	are

19 Find the numbers from one to ten.

o	t	h	r	f	o	u	r
e	h	o	n	t	e	n	x
n	r	v	n	h	f	e	i
i	e	s	c	e	i	g	h
n	e	x	s	e	v	e	n
e	t	w	o	s	n	g	h
f	w	e	s	f	i	v	e
o	e	i	g	h	t	x	r

20 Tick (✓) the correct answer.

Teacher is	a	a nationality	☐
	b	a job	✓
	c	a country	☐

1 *Listen* is	a	classroom language	☐
	b	spelling	☐
	c	grammar	☐

2 *Eight* is	a	classroom language	☐
	b	a number	☐
	c	a job	☐

3 *Thai* is	a	a country	☐
	b	a name	☐
	c	a nationality	☐

4 *Goodbye* is	a	spelling	☐
	b	a greeting	☐
	c	information	☐

5 *Britain* is	a	a nationality	☐
	b	a name	☐
	c	a country	☐

WRITING

21 Write sentences about these people.

Name	Country	Town	Job
Eleni	Greece	Thessaloniki	doctor
Anna	Britain	Manchester	journalist
Marcos	Brazil	Rio de Janeiro	teacher
Murat	Turkey	Ankara	student

Eleni is Greek. She's from

Thessaloniki and she's a doctor.

_____

_____

_____

_____

_____

_____

22 Continue the conversation between Anne and Walter. Ask about names, jobs, nationalities.

Name	Job	Nationality
Walter	Actor	American
Anne	Singer	British

Walter   Hello, _what's your name?_

Anne   I'm Anne._____

_____

Walter   I'm Walter._____

_____

Anne   I'm a singer._____

_____

Walter   _____

_____

Anne   _____

_____

Walter   _____

_____

Lessons 6 -10

GRAMMAR

1 Complete the sentences with *I, he, she* or *they*.

*They*_____ 're my friends.

1 _____'re actors.

2 _____'m from Spain.

3 _____ isn't a secretary.

4 Are _____ your glasses?

5 _____'s my sister.

6 Is _____ your brother?

7 _____ aren't my neighbours.

8 Tom's 19. _____ isn't 20.

9 _____'m not American.

2 Complete the chart with the verb *to be*.

I'm	*I'm not*	Am I?
_____	You aren't	Are you?
_____	_____	Is he?
She's	_____	_____
_____	It isn't	_____
_____	_____	Are we?
They're	_____	_____

3 Write short answers.

Are your neighbours American?

Yes, *they are.*_____

1 Is Louis 18?

No, _____

2 Are John and Anne French?

No, _____

3 Are we friends?

Yes, _____

4 Is this your wallet?

Yes, _____

5 Is Sonia a teacher?

No, _____

6 Are those her cassettes?

No, _____

7 Is Madonna your favourite singer?

Yes, _____

8 Is the car American?

No, _____

9 Are Jack and Mandy married?

Yes, _____

10 Am I a good student?

Yes, _____

4 Complete the passage.

her his he's they're she's is are

Paul's a student. _*He's*_ from New York.

(1) _____ favourite actress

(2) _____ Kim Basinger. Mary's a

teacher. (3) _____ from London.

(4) _____ favourite singers

(5) _____ Michael Jackson and Tina

Turner. (6) _____ American singers.

5 Write the questions and answers. Use *this, that, these* or *those*.

What are these?
They're *sunglasses*

1 _____
It's a _____

2 _____
It's a _____

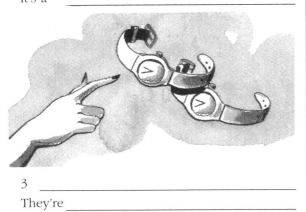

3 _____
They're _____

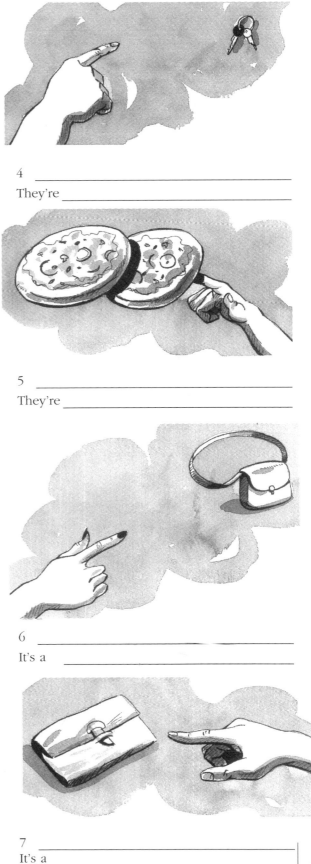

4 _____
They're _____

5 _____
They're _____

6 _____
It's a _____

7 _____
It's a _____

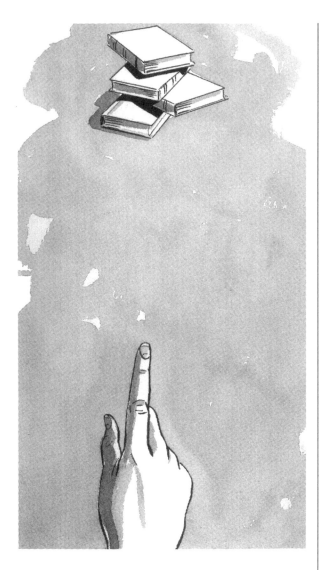

8 _____

They're _____

6 Match the questions and answers.

1 Where are you from?   a Soccer.
2 How old is he?   b No, they
                               aren't.
3 What's his favourite sport? c Tom Cruise.
4 Who's her favourite actor?   d She's a doctor.
5 Are they English?   e Taiwan.
6 Are you married?   f 21.
7 What's her job?   g No, I'm not.

7 Write the negative.

They're my glasses.
They aren't my glasses.

1 We're twins.

_____

2 It's an umbrella.

_____

3 Daniel's from Canada.

_____

4 Albert and Sandy are students.

_____

5 You're 25.

_____

6 His favourite singer is Elton John.

_____

7 They're Japanese.

_____

8 I'm a politician.

_____

8 Write *my* or *your* in the conversations.

A Is this __*your*__ umbrella?
B No, __*my*__ umbrella is in my bag.

1 A How do you do? _____ name's Joy.
  B Hello. _____ name is Fred.

2 A Is she _____ sister?
  B No, she isn't. She's _____ friend.

3 A Are they _____ glasses?
  B Yes, they are. Thank you.

4 A What's _____ job?
  B I'm an engineer.

5 A Is Patricia _____ English teacher?
  B No, she isn't. _____ English teacher is
    Judith.

9 Write *his* or *her*.

What's ___*her*___ phone number?
It's 01675 678093.

1 What's _____ name?
_____ name's Kevin.

2 What's _____ nationality?
She's German.

3 Who's _____ favourite group?
_____ favourite group is Oasis.

4 What's _____ favourite car?
_____ favourite car's a BMW.

5 What's _____ job?
He's an actor.

6 What's _____ address?
It's 12, Park Road, London.

10 Correct the sentences.

What this is? It's my personal stereo.
What's this? It's my personal stereo.

1 Is she teacher? Yes, she is.

_____

2 Are Tom and Gordon doctors? Yes, they're.

_____

3 Are you from Istanbul? No, I amn't.

_____

4 How old you are?

_____

5 Who's your favourite car?

_____

6 What's your favourite actor?

_____

7 The womans are from Paris.

_____

8 What's these? It's a watch.

_____

9 Is Pauline a secretary? No, she is not.

_____

10 She's Spanish. His name's Maria.

_____

11 Write *Who, What, Where* or *How*.

*Who*_____ is Bill? He's my brother.
1 _____ is he? He's Ken.
2 _____ are those? They're cassettes.
3 _____ is your name? It's Faye.
4 _____ is her job? She's a teacher.
5 _____ old are they? They're 26.
6 _____ are you from? I'm from Japan.
7 _____ are you? I'm fine.
8 _____ is Lisbon? It's in Portugal.

SOUNDS

12 Underline the stressed syllables.

<u>Who</u>'s your <u>favourite</u> <u>actress</u>?
1 She's twenty-seven.
2 Jane's American and she's an actress.
3 Is he a student?
4 Her favourite car is a Porsche.
5 What's your address?
6 My English teacher's from Los Angeles.
7 I'm twenty and I'm from England.
8 Is Marc French?

13 Write the words in the correct columns according to their stress.

books   glasses   cassette   watch
pizza   watches   women   names   repeat
address   fifty   boys   fine   translate

□	□□	□□
watch	_____	_____
_____	_____	_____
_____	_____	_____
_____	_____	_____

14 Write the plurals in the correct column. Do they end in /s/ or /z/?

pens   clocks   cassettes   glasses   twins
wallets   friends   umbrellas   televisions
books   students   bags   girls
brothers   groups

/s/	/z/
	pens

VOCABULARY

15 Write the answers in words.

85 - 10 =

seventy-five

1   60 + 20 =

2   5 x 3 =

3   29 - 6 =

4   90 ÷ 2 =

5   11 x 3 =

6   25 + 26 =

7   48 ÷ 2 =

8   70 - 6 =

9   40 + 33 =

10   3 x 4 =

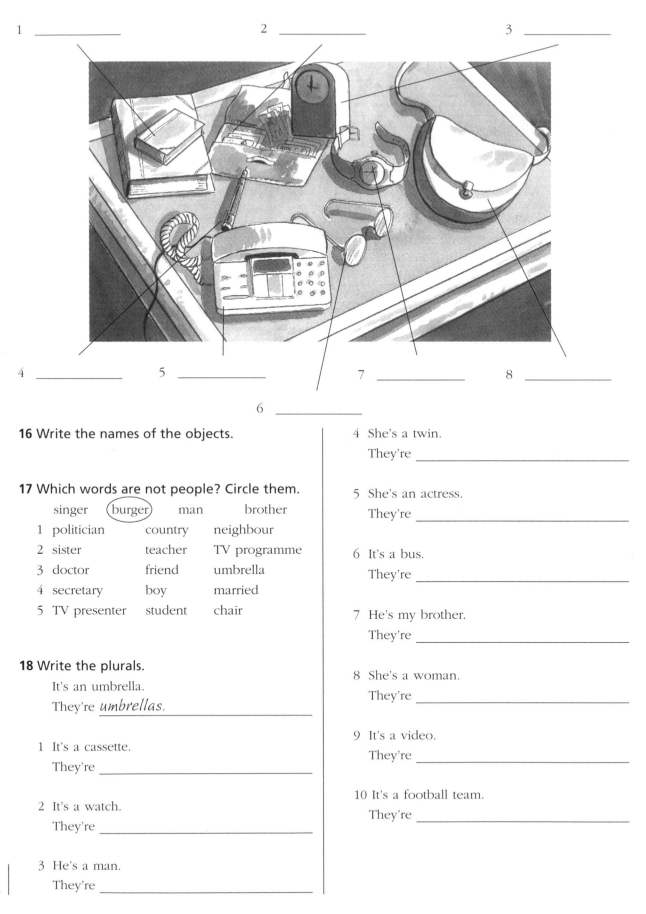

1 _____   2 _____   3 _____

4 _____   5 _____   7 _____   8 _____

6 _____

16 Write the names of the objects.

17 Which words are not people? Circle them.

    singer   (burger)   man   brother

1 politician   country   neighbour

2 sister   teacher   TV programme

3 doctor   friend   umbrella

4 secretary   boy   married

5 TV presenter student   chair

18 Write the plurals.

It's an umbrella.

They're *umbrellas*. _____

1 It's a cassette.

They're _____

2 It's a watch.

They're _____

3 He's a man.

They're _____

4 She's a twin.

They're _____

5 She's an actress.

They're _____

6 It's a bus.

They're _____

7 He's my brother.

They're _____

8 She's a woman.

They're _____

9 It's a video.

They're _____

10 It's a football team.

They're _____

19 Complete the words in the puzzle. Find the hidden word.

1 They're _____ Tokyo.
2 One _____ . Two men.
3 70.
4 _____ old is she?
5 The Spice Girls are an English _____ .
6 She's from Paris. She's _____ .
7 Is he _____ ? No, he's single.
8 Sharon Stone is an _____ .
9 One woman. Two _____ .

Hidden word   _____

20 Match.
1 Job   a brother
2 Age   b Japanese
3 Classroom object   c tennis
4 Sport   d twenty-nine
5 Family   e pen
6 Nationality   f TV presenter

WRITING

21 Complete the questionnaire. Write true answers.

What's your name?

_____

What's your address?

_____

What nationality are you?

_____

How old are you?

_____

Are you married?

_____

What's your job?

_____

What's your favourite TV programme?

_____

Who's your favourite actor?

_____

22 Write a paragraph about yourself. Use the information in Question 21.

_____

_____

_____

_____

_____

_____

_____

Lessons 11-15

GRAMMAR

1 Complete the sentences with *have got*.

We ___*have got*___ nice neighbours.

1 _____ you _____ any brothers and sisters?

2 Marion _____ dark hair.

3 No, I _____ any brothers. I _____ one sister.

4 We _____ one daughter.

5 _____ he _____ blue eyes? Yes, he _____ .

6 No, the twins _____ red hair. They _____ dark hair.

7 _____ Jill _____ any children? No, she _____ .

8 _____ your husband _____ a car? Yes, he _____ .

2 Complete the chart with *have got*.

I've got	*I haven't got*
_____	You haven't got
He's got	_____
She's got	_____
_____	It hasn't got
We've got	_____
_____	They haven't got

Have I got?	Yes, I have.
_____	No, you haven't.
Has he got?	Yes, _____
Has she got?	No, _____
_____	Yes, it has.
_____	No, we haven't.
Have they got?	Yes, _____

3 Complete the sentences with possessive adjectives.

You've got a black car. It's ___*your*___ car.

1 I've got a brother. He's _____ brother.

2 She's got a coat. It's _____ coat.

3 They've got a sister. She's _____ sister.

4 You've got Levi jeans. They're _____ jeans.

5 We've got a son. He's _____ son.

6 He's got red shoes. They're _____ shoes.

4 Match.

1 Open	a pen down
2 Stand	b your jacket on
3 Sit	c door
4 Put your	d your book
5 Pick your	e down
6 Put	f up
7 Come	g bag up
8 Close the	h in

5 Write the negative of the instructions in Question 4.

1 *Don't open your book.*

2 _____

3 _____

4 _____

5 _____

6 _____

7 _____

8 _____

6 Complete the questions and answers. Use *have got*.

Has he got a personal stereo?
No, he hasn't.

1 _____ a watch?

_____

2 _____ long hair?

_____

3 _____ a bag?

_____

4 _____ glasses?

_____

5 _____ fair hair?

_____

6 _____ a wallet in his hand?

_____

7 Write *a, an, the* or *–*.

How much is __*the*__ red shirt?

1 They've got _____ son and _____ daughter.

2 _____ pen on _____ table is red.

3 She's got _____ blue jeans.

4 Is Larry _____ student?

5 Put _____ cassette in _____ cassette player, please.

6 I've got _____ English book.

7 What are they? They're _____ bags.

8 What is it? It's _____ wallet.

9 She's got _____ blue eyes and _____ fair hair.

10 Open _____ window, but don't open _____ door!

8 Write the possessive with *'s*.

Sandra/keys
They're *Sandra's keys.* _____

1 Tim/wallet
It's _____

2 Teresa/brother
He's _____

3 Marina/children
They're _____

4 Paul and Liam/books
They're _____

5 Frank/sweaters
They're _____

6 Anna/skirt
It's _____

7 Ben and Eva/parents
They're _____

8 David/sister
She's _____

9 Manuel/father
He's _____

10 Laura/teacher
She's _____

9 Write *have got* or *be* in the sentences.

Virginia ___*has got*___ fair hair and blue eyes. She _____*is*_____ nice, but very quiet.

1 Dimitri _____ tall. He _____ dark hair and blue eyes. He _____ married.

2 Sara _____ long hair. She _____ very pretty and she _____ friendly, too.

3 Nick and Steve _____ twins. They _____ fair hair and brown eyes. They _____ nice.

4 I _____ small. I _____ short hair and blue eyes. I _____ 44 years old.

5 My brother _____ good-looking. He _____ married. He _____ two children.

6 Her parents _____ very friendly. Her mother _____ 46 and her father _____ 50.

10 Write the questions.

Is she married?
No, she's single.

1 _____
He's 21.

2 _____
I'm fine, thanks.

3 _____
She's tall and friendly.

4 _____
The white jacket's £40.

5 _____
Yes, I've got two brothers.

6 _____
Nancy's bag is under the chair.

7 _____
They're quite good-looking.

8 _____
No, he hasn't got a daughter.

11 Correct the sentences.

How much are the shoes blue?
How much are the blue shoes?

1 They're the Jane's keys.

_____

2 The shoes black are on the chair.

_____

3 How much is these jeans?

_____

4 The greens sweaters are £50.

_____

5 The bag's George is under the table.

_____

6 They got two brothers.

_____

SOUNDS

12 Underline the stressed syllables.

<u>Close</u> your <u>book</u>!

1 How much are the red shirts?
2 They're twenty pounds.
3 Dan's wallet's under the table.
4 Kate's glasses are in her pocket.
5 We've got two sisters and a brother.
6 Have you got any children?
7 He's got dark hair and green eyes.

13 Find the words with the same vowel sound.

wallet   jeans   red   skirt   shoes

jacket   tall   husband   book

blue   *shoes* _____

1 sweater _____

2 bag _____

3 look _____

4 got _____

5 son _____

6 keys _____

7 shirt _____

8 daughter _____

VOCABULARY

14 Write sentences with *in, on* and *under*.

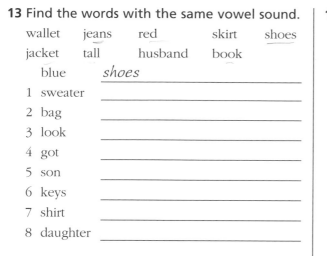

There are some shoes in the sports bag.

1 _____

2 _____

3 _____

4 _____

5 _____

15 Write the questions.

How much are the trainers?

They're £47.50.

1 _____

It's £55.

2 _____

They're £69.

3 _____

They're £39.50.

4 _____

It's £70.

5 _____

They're £20.

21

16 What colour are they?

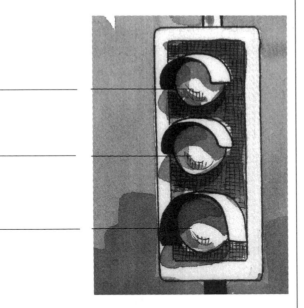

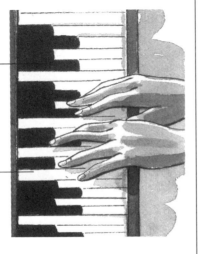

17 Put the letters in the right order and rewrite the passage.

My brother's a tcordo. He's got trsho hair and egrne eyes. He's equti dogo-lkoogni and he's vyre lfyreidn. He's rrmadie and he's got three children: a ons and two ghdaeruts.

My brother's a doctor.

_____

_____

_____

_____

_____

_____

18 Put the words in the right columns.

red   £9.99   jeans   jackets   black   £50
shoes   white   trainers   £45.50   green
50p   £6.80   blue   brown   sweater
£59.60   skirt

Clothes	Colours	Prices
	red	

19 Look at the family tree. Complete the sentences with the words below.

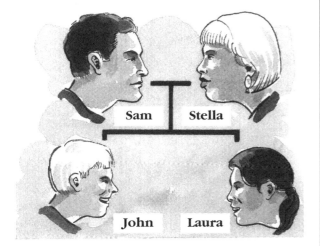

Sam **Stella**

John **Laura**

daughter   father   husband   mother
brother   son   sister   wife
children   parents

John is Laura's *brother*. _____

1 Sam is Stella's _____ .
2 Stella is Laura and John's _____ .
3 John is Sam and Stella's _____ .
4 Laura is John's _____ .
5 Sam is Laura and John's _____ .
6 Laura is Stella and Sam's _____ .
7 John and Laura are Stella and Sam's

_____ .

8 Stella is Sam's _____ .
9 Stella and Sam are John and Laura's

_____ .

20 Circle the odd-one-out.

watch   keys   (mother)   glasses   wallet
1 read turn pen put go
2 red green blue jeans white
3 black father children sister daughter
4 skirt shirt coat jacket light
5 chair come table window door
6 friendly good-looking pretty quiet very

WRITING

21 Fill in the chart for two friends. Describe them.

Name		
Age		
Appearance		
Job		
Town		

_____

_____

_____

_____

_____

_____

_____

_____

22 Describe yourself.

_____

_____

_____

_____

_____

_____

GRAMMAR

1 Write *do, are* or *have* in the sentences.

What time ___*do*___ you have breakfast?

1 _____ you from China?

2 _____ you have lunch with your family?

3 My parents _____ got a house in Boston.

4 You _____ in my English class.

5 _____ you like shopping?

6 We _____ dinner in a restaurant on Saturdays.

7 _____ you work in an office?

8 The boys _____ at school.

9 _____ you got a flat or a house?

10 _____ they go to school on Wednesday?

2 Complete the chart with the present simple.

I work   *I don't work*

_____   You don't write

We have   _____

They live   _____

(I/work)   *Do I work?*

Yes, *I do.*_____

(you/write) _____

No, _____

(we/have) _____

Yes, _____

(they/live) _____

No, _____

3 Write the third person.

I work. She *works*._____

1 I live. She _____

2 I work. He _____

3 I leave. He _____

4 I go. She _____

5 I have. It _____

6 I start. He _____

7 I finish. She _____

8 I arrive. It _____

4 Complete the passage.

works   has   watches   goes   finishes
lives   arrives   visits   is   starts
corrects   has   leaves

Miriam ___*lives*___ in a house in Dallas.

She (1) _____ in the mornings in a school. She (2) _____ a teacher. She (3) _____ breakfast at 7.30 and (4) _____ the house at 8. She (5) _____ at school at 8.45 and (6) _____ work at 9. At 11 o'clock she (7) _____ coffee with her colleagues. She (8) _____ work at 12.30. In the afternoon, she (9) _____ shopping and (10) _____ friends. In the evening, she (11) _____ homework and (12) _____ television.

5 **Is the 's'** *is, has* **or possessive?**

I've got Andrea's passport.
possessive _____

1 They're Mr Kane's keys.

_____

2 Pierre's married.

_____

3 Eva's got a flat in Toronto.

_____

4 The President's in London.

_____

5 Her husband's name's Bruno.

_____

6 She's Murat's daughter.

_____

7 He's got an English class on Monday.

_____

8 He's my friend's brother.

_____

9 It's a German car and it's blue.

_____

10 It's Paul's house.

_____

6 **Write the negative.**

They live in a flat.
We *don't live in a flat.* _____

1 We like football.
They_____

2 George's got a brother.
Mike_____

3 Julie's married.
Karen_____

4 They're Turkish.
We _____

5 We've got dark hair.
They_____

6 You work in Boston.
I _____

7 I have dinner at 9 pm.
You _____

8 His name's Bill Bailey.
My name _____

7 **Write** *in, on, at* **or** *to.*

My brother lives _*in*_ Atlanta _*in*_ the USA.

1 We go _____ school _____ London.

2 Paula lives _____ a flat _____ Rome.

3 I arrive _____ work _____ 9 am.

4 Maria visits her parents _____ Sundays.

5 Do you have breakfast _____ 7 _____ the morning?

6 The girls listen _____ music _____ the evening.

7 He goes _____ the cinema _____ Saturday evening.

8 **Write short answers.**

Is he an engineer?
Yes, *he is.* _____

1 Do you live in Istanbul?
Yes, _____

2 Do we have English tomorrow?
No, _____

3 Do they like skiing?
Yes, _____

4 Has she got fair hair?
Yes, _____

5 Are they from Thailand?
No, _____

6 Do you work in an office?
Yes, _____

7 Have you got a brother?
No, _____

9 Complete the questions.

Do you live _____ in a house?
No, I don't.

1 _____ sailing?
Yes, I do.

2 _____ time?
It's 4 o'clock.

3 _____ married?
Yes, I am.

4 _____ breakfast at 5 o'clock?
No, I don't.

5 _____ shopping on Fridays?
Yes, he does.

6 Where _____ ?
She's from Korea.

7 _____ a flat?
Yes, I have.

10 Complete the sentences with *have got* or *have*.

He _____ *has* _____ lunch with his friend.

1 I _____ dinner at 8 pm.

2 We _____ a new neighbour.

3 She _____ long hair and blue eyes.

4 They _____ breakfast at 10 am.

5 _____ you _____ any children?

11 Correct the mistakes.

You have breakfast at 8.30?
Do you have breakfast at 8.30?

1 We go shopping on the morning.

_____

2 Her husband work in New York.

_____

3 I go at home at 7 pm.

_____

4 I like very much basketball.

_____

5 They've got lunch at 1 pm.

_____

6 We work not on Saturday and Sunday.

_____

SOUNDS

12 Underline the stressed syllables.

I <u>vi</u>sit my <u>par</u>ents on <u>Sun</u>days.

1 I work in an office in Madrid.

2 His children go to school in the morning.

3 Do you like basketball?

4 I like gymnastics very much.

5 Do you work in a school?

6 What's the time, please?

7 They live in a house in Chicago.

13 Find words from Units 16-20 to write in these columns.

▢▢   ▢▢▢   ▢▢   ▢▢▢

_____ _____ _____ _____

_____ _____ _____ _____

_____ _____ _____ _____

_____ _____ _____ _____

_____ _____ _____ _____

_____ _____ _____ _____

_____ _____ _____ _____

_____ _____ _____ _____

_____ _____ _____ _____

_____ _____ _____ _____

Which columns have got the most words?

_____

VOCABULARY

14 Put the phrases into the correct column.

5.30   Sunday morning   the evening
three o'clock   Wednesday
the afternoon   Thursday evening
a quarter past two   the morning
Sundays

on	in	at
		at 5.30

15 Match the times with the activities.

1 7.30 am   a finish work
2 8 am   b start work
3 8.30 am   c have breakfast
4 1 pm   d arrive home
5 5.30 pm   e leave home
6 6 pm   f have lunch

16 Which sports do you play with a ball?

tennis

Day	Morning	Afternoon	Evening
Monday			

17 Complete the chart with the days of the week. Then write the things you do.

18 Match the words.

1 watch   a   shopping
2 play   b   lunch
3 live   c   in an office
4 go   d   in a flat
5 work   e   television
6 have   f   a newspaper
7 read   g   to music
8 listen   h   volleyball

19 Put the times in order.

a   quarter past five in the afternoon ☐
b   half past eleven in the morning ☐
c   quarter to ten in the evening ☐
d   one o'clock in the afternoon ☐
e   quarter past nine in the morning ☐
f   eleven o'clock in the evening ☐
g   half past eight in the evening ☐
h   quarter to twelve in the morning ☐

20 Write true sentences.

1 I like _____ , _____ ,
and _____ .
2 I don't like _____ ,
_____ , and _____ .

3 My friend likes _____ ,
_____ , and _____ .
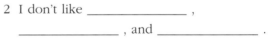
4 My teacher likes _____ ,
_____ , and _____ .

Name	Wayne	Fiona	Mick and May	You
Nationality	American	Scottish	Australian	
Job	singer	shop assistant	students	
Family	2 sisters	married, no children	twins	
Home	flat	house	at home with parents	
Hobbies	basketball, skiing	reading	sailing	

WRITING

21 Complete the chart and write sentences.

Wayne is American. He's _____

_____

Fiona _____

_____

Mick and May _____

_____

I'm _____

_____

_____

22 Look at the chart in Question 17 and write a paragraph about what you do in a week.

_____

_____

_____

_____

_____

_____

_____

Lessons 21-25

GRAMMAR

1 Complete the sentences.

What ___do___ you usually do on your birthday?

1 _____ Michel work in France?

2 No, I _____ like writing letters.

3 What _____ you have for breakfast?

4 There _____ five tables in the classroom.

5 No, she _____ stay in hotels.

6 _____ there a sofa in the living room?

7 _____ you got a car?

8 They _____ waiters.

9 No, there _____ any milk in the fridge.

10 When _____ it leave?

2 Complete the chart with the present simple.

I eat	_I don't eat._
_____	You don't drink
He likes	_____
She lives	_____
_____	It doesn't have
_____	We don't walk
They go	_____

(I/eat)   _Do I eat?_

Yes, _I do._

(you/drink) _____

No, _____

(he/like) _____

Yes, _____

(she/live) _____

No, _____

(it/have) _____

Yes, _____

(we/walk) _____

No, _____

(they/go) _____

Yes, _____

3 Write the short answers.

Do you walk to work?

No, _I don't._

1 Does she sometimes go to work by bus?

Yes, _____

2 Do your parents like sightseeing?

No, _____

3 Is there a cooker in the classroom?

No, _____

4 Has your teacher got blue eyes?

Yes, _____

5 Are there any students in the restaurant?

Yes, _____

6 Does Mohammed live in Morocco?

Yes, _____

7 Do you like lying on the beach?

No, _____

8 Are Tom and Dave from Manchester?

No, _____

9 Does the train always leave at 7 am?

Yes, _____

10 Do you spend your birthday with friends?

Yes, _____

4 Write sentences from the information.

	Patrick	**Mary**	**Joe**
dancing	✗	✓	✓
reading	✓	✗	✓
having parties	✗	✓	✓
doing homework	✓	✗	✗

Patrick doesn't like dancing.

1 _____

2 _____

3 _____

4 _____

5 _____

6 _____

7 _____

5 Write the sentences in the correct order.

We/parties/don't/have/often
We don't often have parties.

1 sightseeing/My/often/parents/go

_____

2 have/Mondays/We/on/always/English

_____

3 do/do/in/What/usually/you/August

_____

4 sometimes/Paul/friends/home/invites

_____

5 letters/never/I/write

_____

6 Match.

1 There's   a any plants in the kitchen?

2 Are there   b any vegetables, but
                          there's lots of fruit.

3 There isn't   c a television in the living
                          room.

4 Is there   d a bath, but there's a
                          shower.

5 There aren't   e four bedrooms in my
                          house.

6 There are   f a bathroom upstairs?

7 Write questions.

What do you have for breakfast?
I have coffee and bread for breakfast.

1 _____

Neil goes to work by train.

2 _____

Alexandra starts work at 9 am.

3 _____

They live in Bangkok.

4 _____

There's a tennis racquet on the table.

5 _____

It's £4.99.

6 _____

She's 19 years old.

7 _____

We like swimming and walking.

8 What do you do on your birthday? Write true sentences. Use the adverbs in brackets.

1 (always)

_____

_____

2 (usually)

_____

_____

3 (often)

_____

_____

4 (sometimes)

_____

_____

5 (never)

_____

_____

9 Match.

1 Does she sometimes   a   lying on the beach.
2 We   b   go to school by taxi?
3 They don't   c   from Japan.
4 I like   d   never eat meat.
5 He's   e   often read.
6 They've   f   got three children.

10 Write true answers.

1 Where do you live?

_____

2 Do you go to school/work by bus?

_____

3 What time do you have lunch?

_____

4 What do you usually have for lunch?

_____

5 Who do you spend your birthday with?

_____

6 Do you get any cards for your birthday?

_____

7 What do you often do at New Year?

_____

11 Choose the correct answer.

   a Often I see my parents on Sunday. ☐
   b I see often my parents on Sunday. ☐
   c I often see my parents on Sunday. ☑
1 a There is lots of food in the fridge. ☐
   b There are lots of food in the fridge. ☐
   c There has lots of food in the fridge. ☐
2 a He don't like sitting in the sun. ☐
   b He doesn't like sit in the sun. ☐
   c He doesn't like sitting in the sun. ☐
3 a What eat you for dinner? ☐
   b What do you eat for dinner? ☐
   c What you do eat for dinner? ☐
4 a She go to work at 8.30. ☐
   b She goes to work at 8.30. ☐
   c She does go to work at 8.30. ☐
5 a There's a living room downstairs. ☐
   b It's a living room downstairs. ☐
   c This is a living room downstairs. ☐
6 a Sophie doesn't never drink coffee. ☐
   b Sophie drinks never coffee. ☐
   c Sophie never drinks coffee. ☐
7 a Do you like swimming? ☐
   b Do you like swim? ☐

SOUNDS

12 Underline the stressed syllables.

Do you like lying on the beach?

1 Max goes to school by car.

2 When's your birthday?

3 I often go to restaurants.

4 What do you do with your friends?

5 There's a cupboard in the hall.

6 Are there any people in the classroom?

7 I like staying in hotels.

13 Which words have the same vowel sound?

on   go   do   fourth   out

cooker   word   front   pocket

watch   *pocket*

1 look   _____

2 for   _____

3 don't   _____

4 does   _____

5 to   _____

6 down   _____

7 often   _____

8 work   _____

VOCABULARY

14 Circle the odd-one-out.

sea	beach	sun	(work)
1 month	yoghurt	day	week
2 July	May	Friday	June
3 twelve	tenth	third	second
4 bread	rice	potato	beer
5 eating	walking	swimming	skiing

15 Write words.

Snacks
sandwich, pizza

1 Drinks you have in the morning

_____

_____

2 Meats

_____

_____

3 Drinks you have with main meals

_____

_____

4 Fruit

_____

_____

5 Vegetables

_____

_____

6 Food you eat for breakfast

_____

_____

7 Your favourite food

_____

_____

16 Put the letters in the correct order and rewrite the passage.

In my house, there's a gdinin room, a large viglni room, a ickhnet and a llah downstairs. sarpsUti, there are three sdomrobe and a mobotahr. My room is at the kbca and there is a big dnwowi onto the rnedga.

In my house, there's a dining room,

_____

_____

_____

_____

_____

17 Write the dates of these famous days.

Bastille Day
14th July

1 Valentine's Day

_____

2 US Independence Day

_____

3 New Year's Day

_____

4 Christmas Day

_____

5 Your birthday

_____

6 An important day in your country

_____

18 Complete the 'holiday' puzzle. Find the hidden word.

1 I like _____ in restaurants.
2 We always _____ coffee in Italy.
3 I like lying on the _____ .
4 We sometimes _____ paella for dinner in Spain.
5 We don't like staying in _____ .
6 We usually go on holiday in June or _____ .
7 My husband _____ swimming in the sea.
8 We like _____ in the disco in the evening.
9 On holiday I like sitting in the _____ .
10 In winter we like _____ in the mountains.

Hidden word   _____

19 Write words.

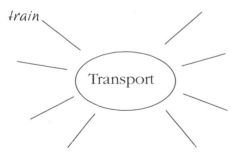

train

Transport

20 Tick (✓) the correct boxes.

You like

a lying on the beach. ✓

b reading. ✓

c work by bus. ☐

1 You eat

a meat. ☐

b tea. ☐

c vegetables. ☐

2 You get

a presents. ☐

b postcards. ☐

c swimming. ☐

3 You drink

a apple pie. ☐

b apple juice. ☐

c apples. ☐

4 You have

a lunch. ☐

b a party. ☐

c 19 years old. ☐

5 You go

a to holiday. ☐

b at work. ☐

c shopping. ☐

WRITING

21 Describe the house or flat where you live.

_____

_____

_____

_____

_____

_____

22 Describe breakfast, lunch and dinner in your country.

Breakfast

_____

_____

_____

Lunch

_____

_____

_____

Dinner

_____

_____

_____

23 What do you like doing on holiday?

_____

_____

_____

_____

Lessons 26-30

GRAMMAR

1 What can/can't they do? Write sentences.

	John	**Kathy**
cook	✓	✗
speak Spanish	✗	✓
drive	✓	✓
play the guitar	✓	✗
draw well	✗	✓

John can cook.

1 _____

2 _____

3 _____

4 _____

5 _____

6 _____

7 _____

8 _____

9 _____

2 Complete the job interview.

Interviewer *Are you* a secretary?

Tina No, I _____ . I'm a student.

Interviewer So you want a holiday job in an office?

Tina Yes, I _____ .

Interviewer Well, _____ type?

Tina Yes, I _____ . And I _____ use a computer.

Interviewer _____ speak French?

Tina No, I _____ , but I _____ German.

Interviewer Well, _____ start tomorrow?

Tina Yes, I _____. Thank you!

3 Complete the chart with the present continuous.

I'm making *I'm not making*

You're having _____

_____ He isn't reading

She's waiting _____

It's working _____

_____ We aren't talking

They're listening _____

(I/make) *Am I making?*

Yes, *I am.*

(you/have) _____

No, _____

(he/read) _____

Yes, _____

(she/wait) _____

No, _____

(it/work) _____

Yes, _____

(we/talk) _____

No, _____

(they/listen) _____

Yes, _____

4 Write the present participle form.

play _____playing_____

sit _____

run _____

drink _____

lie _____

write _____

have _____

stand _____

do _____

5 Write the sentences in the correct order.

you/are/Here

Here you are. _____

1 I'd/a/coffee/like/of/cup

_____

2 to/Anything/eat/?

_____

3 very/Thank/much/you

_____

4 you/Can/help/I/?

_____

5 I/have/sandwich/Can/a/cheese/too/?

_____

6 Write a conversation in a restaurant with the sentences from Question 5.

Waitress _Can I help you?_____

Customer _____

_____, please.

Waitress _____

Customer Yes._____

Waitress Certainly. _____

Customer ___ _____

7 Write short answers.

Are you cooking my lunch?

No, _I'm not._____

1 Can you swim?

Yes, _____

2 Are they having a sandwich?

No, _____

3 Are the students listening to the teacher?

Yes, _____

4 Can Martin sing well?

No, _____

5 Is Jilly talking to her sister?

Yes, _____

6 Do you like swimming?

Yes, _____

7 Have your parents got a big car?

No, _____

8 Are there any computers in the classroom?

No, _____

9 Does Bill live in England?

Yes, _____

10 Can they speak Japanese?

Yes, _____

8 Tick (✓)the correct answer.

They watch TV

a this evening. ☐

b in the evening. ☑

1 He's having a bath

a now. ☐

b every day. ☐

2 Every morning at 7 o'clock

a she has breakfast. ☐

b she's having breakfast. ☐

3 She uses a computer

a at work. ☐

b this afternoon. ☐

4 Look at Fred!

a He drives his new car. ☐

b He's driving his new car. ☐

5 I'm waiting for the bus

a at the moment. ☐

b every morning. ☐

6 I play tennis

a on Saturdays. ☐

b this Saturday. ☐

7 It's 1 o'clock in the morning.

a Most people sleep. ☐

b Most people are sleeping. ☐

9 What's the difference between the present simple and the present continuous? Write in your language if you prefer.

_____

_____

_____

_____

10 Write questions.

*What's he playing?*_____

He's playing golf.

1 _____

Robin's from Atlanta.

2 _____

They're sitting in a cinema.

3 _____

I'm cooking pasta.

4 _____

No, I can't play the piano.

5 _____

I live in Toronto.

6 _____

She wants a computer for her birthday.

7 _____

Yes, she does. She likes music very much.

11 Correct the mistakes.

Are you understanding English?

Do you understand English?

1 I standing in a queue.

2 You are buying lunch?

3 The men is sitting in a bar.

4 She isn't wanting a sandwich at the moment.

5 My mother is shoping this morning.

6 Have you need help?

7 I can't speaking Russian.

SOUNDS

12 Underline the stressed syllables.

Can you speak <u>English</u>?
1 I can speak French, but I can't speak Italian.
2 What about sport?
3 Can I help you?
4 Anything to drink?
5 A piece of chocolate cake, please.
6 Where's the library, please?
7 Turn right at the bank and go straight ahead.
8 They're getting up and having breakfast.
9 What are you doing at the moment?

13 Which vowel sound is different? Circle the odd-one-out.

	speak	piece	(live)	street
1	right	type	pie	swim
2	east	French	bread	left
3	straight	bank	baked	cake
4	lie	sit	in	drink
5	car	can't	can	bar

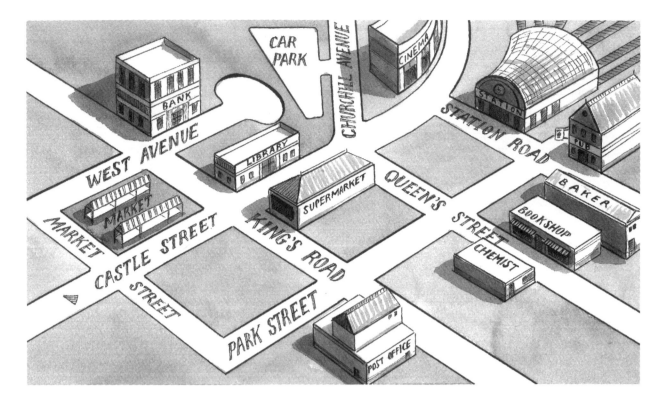

VOCABULARY

14 Look at the street plan. Write directions to these places.

Where's the cinema?
Go along Castle Street. Go past
Churchill Avenue. The cinema is on
your left.

1 Where's the post office?

_____

_____

_____

_____

2 Where's the car park?

_____

_____

_____

_____

3 Where's the bookshop?

_____

_____

_____

_____

_____

_____

15 Look at the street plan. Follow the directions. Where are you?

1 Turn right into Market Street. Turn left into Park Street and go straight ahead. Go past Station Road and it's on your right.
You are at _____

2 Go straight along Castle Street. Turn right into Station Road and it's on your left.
You are at _____

3 Turn left into Market Street and turn right into West Avenue. Go past King's Road and it's on your left.
You are at _____

16 Put the food and drink on the right menu.

milk pasta tuna and mayonnaise
yoghurt cheese and tomato
chocolate cake coffee beef with rice
cola ice cream chicken and lettuce
pie with vegetables tea cheese
apple pie pizza

17 Complete the sentences.

Can I have a *cup* of coffee, please?

1 I'd like a _____ of cake.

2 Can I have a _____ of wine?

3 A _____ of mineral water, please.

4 I'd like a _____ of tea, please.

18 Complete the sentences for you.

At seven o'clock I _____

_____

_____

1 At half past seven I _____

_____

_____

2 For breakfast I _____

_____

_____

3 For lunch I _____

_____

_____

4 After school/work I _____

_____

_____

5 In the evening I _____

_____

_____

19 Choose a verb and a noun and write true sentences.

Verbs

make   speak   play   use
ride   understand   drive

Nouns

piano   computer   English
apple pie   tennis   car   horse

I can't drive a car. _____

_____

_____

_____

_____

_____

_____

_____

20 Correct the sentences.

You have dinner in the chemist's.
You have dinner in a restaurant. _____

1 You watch a film in the bank.

_____

2 You have a bath in the living room.

_____

3 You buy books in the restaurant.

_____

4 You park your car in the cinema.

_____

5 You take the train at the post office.

_____

WRITING

21 What do you do at the weekend? Write a paragraph. Try to use some of these words: *often, sometimes, usually, always, never.*

_____

_____

_____

_____

_____

_____

_____

22 What are they doing? Look at the picture. Write sentences.

Gordon is eating a sandwich and

reading a book.

_____

_____

_____

_____

_____

_____

Lessons 31-35

GRAMMAR

1 Match the phrases.

1	Let's	a	I don't like the theatre.
2	I'm sorry, but	b	it on?
3	What's	c	go to the theatre.
4	How about	d	on?
5	Where's	e	going to the cinema?

2 Use the phrases in Question 1 to complete the conversation.

Amanda   *Let's go to the theatre* this evening.

Tony   Oh no!_____

_____

Amanda   Well, _____

_____ then?

Tony   _____

Amanda   The new Tom Cruise film.

Tony   And_____

Amanda   At the Odeon Cinema.

Tony   OK, then.

3 Look at Joanna's diary. Write sentences about next week.

○ Go to travel agent's	MONDAY
○	TUESDAY
○ Have lunch with Larry	
○	
Go shopping with Sarah	WEDNESDAY
Get tickets from travel agent's	THURSDAY
Visit parents	FRIDAY
○	
○	SATURDAY
Fly to New York	
○	
	SUNDAY

On Monday, Joanna's going to the

travel agent's.

1 _____

_____

2 _____

_____

3 _____

_____

4 _____

_____

5 _____

4 Tick (✓) the passages which are about future plans.

1 We are sitting in the classroom. A student is talking about her holidays and we are listening to her. ☐

2 Her plane is leaving at 4.30 pm on Monday. She's going to Vancouver. ☐

3 Maria's mother is in the kitchen. She's making a cake for Maria's birthday party. ☐

4 There are a lot of people in the Italian restaurant. They're eating pizza and drinking Chianti wine. They're talking and enjoying their meal. ☐

5 We're having a special birthday meal this evening. We're going to a Moroccan restaurant. We're meeting our friends at 9 pm. ☐

6 My brother and his wife are going to Spain for a week. They're staying in Barcelona for two days, then they're going to Madrid. ☐

5 Write questions from the passages in Question 4.

What is the student talking about?
She's talking about her holidays.

1 _____
It's leaving at 4.30 pm next Monday.

2 _____
She's making a cake.

3 _____
They're talking and enjoying their meal.

4 _____
We're going to a Moroccan restaurant.

5 _____
They're going to Spain.

6 Complete the chart with the past tense of *to be*.

I was	I wasn't
Was I?	Yes, I was
You were	_____
Were you?	No, _____
_____	She wasn't
_____	Yes, _____
We were	_____
_____	No, _____
_____	They weren't
_____	Yes, _____

7 Complete with the past tense of *to be*.

It *was* very quiet in the house.

1 I _____ with my friends yesterday.

2 No, they _____ in the kitchen.

3 _____ you in London yesterday?

4 Professor Swan _____ in the dining room.

5 There _____ a lot of people in the bus.

6 What time _____ it?

7 She _____ hungry, but she _____ thirsty.

8 _____ there a knife on the table?

9 We _____ bored at the concert last night.

8 Tick (✓) the correct word.

Are you coming with   a   us   [✓]

                                 b   our   [ ]

                                 c   we   [ ]

1   I don't like   a   he   [ ]

                       b   him   [ ]

                       c   his   [ ]

2   She's talking to   a   their   [ ]

                          b   they   [ ]

                          c   them   [ ]

3   I was at the theatre with   a   her   [ ]

                                       b   she   [ ]

4   There was a knife on   a   its   [ ]

                               b   it   [ ]

5   He often has lunch with   a   I   [ ]

                                 b   me   [ ]

                                 c   my   [ ]

6   Can I ask   a   you   [ ]

                 b   your   [ ]

                 c   their   [ ]

9 Write the past tense of *to have* in the chart.

I had	*I didn't have*
_____	Yes, I did.
You had	_____
Did you have?	No, _____
_____	He didn't have
_____	Yes, _____
We had	_____
Did they have?	Yes, _____

10 Write questions and answers about Lady Scarlet's living room.

sofa (✓)
Did it have a sofa? Yes, it did.

1   television (✓)

_____

2   video recorder (✗)

_____

3   fax machine (✗)

_____

4   table (✓)

_____

5   telephone (✓)

_____

6   computer (✗)

_____

11 Put the verbs into the correct tense.

We (not have) a holiday last year.
We didn't have a holiday last year.

1   Where (you, go) next week?

_____

2   I (be) in Paris yesterday.

_____

3   She (not have) breakfast at home every morning.

_____

4   We (be) tired last night.

_____

5   My sister (like) pop music.

_____

6 Her teacher (not work) on Fridays.

_____

_____

7 The boys (have) a party last Saturday.

_____

_____

8 My parents (get) the plane next Sunday.

_____

_____

SOUNDS

12 Underline the stressed syllables.

We <u>didn't</u> have a <u>car</u>.

1 Was she in the kitchen?
2 No, she wasn't.
3 Let's go to the theatre.
4 I'm sorry, but I don't like films.
5 Where's it on?
6 The film was awful.
7 The radio programmes were good.

13 Tick (✓) the words with the stress on the first syllable.

computer	radio✓	dishwasher
bicycle	machine	grandparents
cameras	families	garden
thirsty	butler	alone
unhappy	happy	yesterday
exhibition	theatre	potato
anyone	today	

VOCABULARY

14 Complete the sentences.

getting	packing	meeting
spending	camping	flying
wearing	arriving	staying

Helen's _staying_ in a three-star hotel.

1 I'm _____ my cases.

2 We're _____ our friends in the restaurant at 8 pm.

3 Brian and Jack are _____ in the outback.

4 It's hot! I'm _____ my hat.

5 He's _____ his plane tickets from the travel agent's.

6 Sally is _____ a week in Australia.

7 She's _____ to Sydney.

8 Mr Dayton's _____ at the airport this morning.

15 What do you have in the place where you live? Put a tick (✓).

mountains the sea beaches towns
cinemas theatres galleries restaurants
night clubs an airport railway stations
hotels monuments
Other

_____

16 Write adjectives.

How are you? I'm *fine.*

1 The film wasn't interesting. We were
_____ .

2 In Spain in August, it's _____ .

3 There was no food. They were
_____ .

4 There is no water. We're _____ .

5 _____ birthday to you!

6 The window is open. We're
_____ .

7 At 11.30 pm I'm _____ .

17 What did your grandparents have? What didn't they have? Write four sentences.

My grandparents had a television, but they didn't have a fax machine.

_____

_____

_____

_____

_____

_____

18 Match.

1 You see a film   a in a gallery
2 You look at paintings   b in a club
3 You see a play   c in a cinema
4 You dance   d in a concert hall
5 You listen to music   e in a theatre

19 Put the times in chronological order.

a this morning ☐
b tomorrow ☐
c last year ☐
d last week ☐
e next year ☐
f tonight ☐
g this afternoon ☐
h yesterday ☐
i next week ☐

20 Find six words for each group.

Verbs	Nouns
go	_____
_____	_____
_____	_____
_____	_____
_____	_____
_____	_____

Adjectives	Prepositions	Pronouns
_____	_____	_____
_____	_____	_____
_____	_____	_____
_____	_____	_____
_____	_____	_____
_____	_____	_____

WRITING

21 Where were you? Write true sentences.

Where were you last summer?

_____

_____

Where were you last night at 10 o'clock?

_____

_____

Where were you yesterday morning?

_____

_____

Where were you last weekend?

_____

_____

22 Write the plans for your next holiday.

_____

_____

_____

_____

_____

_____

_____

_____

_____

_____

Lessons 36-40

GRAMMAR

1 Complete the chart with the correct verb forms.

infinitive	-ing	past tense
look	_looking_	_looked_
have	_____	_____
buy	_____	_____
get	_____	_____
come	_____	_____
study	_____	_____
take	_____	_____
stop	_____	_____
go	_____	_____
dance	_____	_____
leave	_____	_____

2 Complete the chart with the past simple.

I stayed	_I didn't stay_
_____	You didn't fly
_____	He didn't eat
She saw	_____
_____	We didn't find
_____	They didn't read

(I stay) _Did I stay?_
Yes, _I did._
(you/fly) _____
No, _____
(he/eat) _____
Yes, _____
(she/see) _____
No, _____
(we/find) _____
Yes, _____
(they/read) _____
No, _____

3 Complete the passage. Put the verbs in the past tense.

arrive	have	see	go
come	visit	be	fly
buy	take	spend	

Lynn and Paul were in Paris last weekend.
They _____flew_____ from London and
(1) _____ at the airport at 10 am
on Saturday morning.
They (2) _____ a taxi from the
airport to their hotel in the centre of Paris. On
Saturday afternoon, they (3) _____
shopping and (4) _____ a lot of
souvenirs. In the evening they
(5) _____ dinner in a French
restaurant. On Sunday they
(6) _____ the Louvre Museum
where they (7) _____ the Mona
Lisa. They (8) _____ three hours in
the museum and they (9) _____
very tired. They (10) _____ home
on Sunday evening.

4 Write questions from the passage.
Where were Lynn and Paul?
They were in Paris.
1 _____
They flew from London.
2 _____
They arrived at 10 am on Saturday morning.
3 _____
They took a taxi.
4 _____
They went shopping.
5 _____
They bought a lot of souvenirs.
6 _____
They went to the Louvre on Sunday.
7 _____
They spent three hours there.

5 Complete the sentences.

| ago | last | every | next |
| this | today | yesterday |

The students went shopping *yesterday.*

1 Sheila lived in Tokyo 3 years _____.

2 I'm having a birthday party _____ evening.

3 We had a great holiday in America _____ year.

4 We go to our English class _____ morning.

5 Mark's studying Maths _____ year at University.

6 She's taking the plane _____ .

6 Write short answers.

Are you going to a show tonight?
No, *I'm not.*

1 Did you take any photographs?
Yes, _____

2 Did Olga have lunch with you?
No, _____

3 Were you late?
No, _____

4 Was the weather good?
Yes, _____

5 Did they get lost?
No, _____

6 Does Virginia live in Scotland?
Yes, _____

7 Have his parents got a big car?
Yes, _____

7 Complete the grammar puzzle. Look at the Grammar Review in your Student's Book to help you.

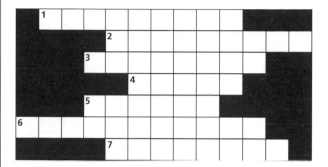

1 *Go, see* and *fly* are _____ verbs.

2 *Always* and *never* are _____ adverbs.

3 *Any* and *a lot of* are expressions of _____ .

4 *Can* is a _____ verb.

5 *Do* and *does* are used in the present _____.

6 *Stand up!* and *Sit down!* are _____ .

7 *A, an* and *the* are _____ .

8 Write sentences about Rick in the negative.

Luigi is Italian.
Rick isn't Italian.

1 Luigi was born in Milan.

_____

2 Luigi went to college in Rome.

_____

3 Luigi likes sport.

_____

4 Luigi drives a Fiat.

_____

5 Luigi's got a big family.

_____

9 Write the verb in the correct tense.

He (have) _____ *has* _____ lunch with his brother every Friday.

1 How much (you, spend) _____ in the shops yesterday?

2 Look! The children (play) _____ volleyball.

3 I (not, sleep) _____ well last night.

4 The weather (be) _____ awful last week.

5 (you, often, go) _____ out with friends?

6 Picasso (live) _____ in France.

7 What (you, do) _____ next summer?

8 Stefan (like) _____ listening to the radio in the evenings.

10 Put the verbs from Question 9 in the correct column.

present simple	present continuous	past simple
*has*	_____	_____
_____	_____	_____
_____	_____	_____
_____	_____	_____
_____	_____	_____
_____	_____	_____
_____	_____	_____
_____	_____	_____
_____	_____	_____
_____	_____	_____
_____	_____	_____

11 Write true answers.

1 Where were you born?

_____

2 Where did you go to school?

_____

3 What sports did you play at school?

_____

4 What sports do you play now?

_____

5 When was the last time you went on holiday?

_____

6 Where did you go?

_____

7 How long did you stay there?

_____

8 Where are you going for your next holiday?

_____

SOUNDS

12 Underline the stressed syllables.

We had a <u>walk</u> in the <u>coun</u>try.

1 He went to London.

2 We left our hotel at 10 o'clock.

3 What did they do in the afternoon?

4 How much were the tickets?

5 She found the cat in her garden.

6 We went home in the evening.

7 When he was fifteen, he left Malaga.

13 Do these past tenses end in /t/, /d/ or /ɪd/?

watched	stayed	wanted	travelled
worked	played	visited	cooked
started	listened	lived	danced
painted	died	called	finished
happened			

/t/	/d/	/ɪd/
watched	_stayed_	_wanted_
_____	_____	_____
_____	_____	_____
_____	_____	_____
_____	_____	_____
_____	_____	_____
_____	_____	_____
_____	_____	_____
_____	_____	

VOCABULARY

14 Write compound nouns.

Lunch you have on Sunday.

Sunday lunch

1 The centre of the city.

_____

2 Games you play on the computer.

_____

3 A sport you play in teams.

_____

4 A room where you wait.

_____

5 An office where you buy tickets.

_____

6 The room where you have dinner.

_____

7 A machine that washes dishes.

_____

15 Put the information in the correct place on the ticket.

DATE:	**BARBICAN HALL**
TIME:	Jazz Scenes
PRICE:	Circle D10
SEAT:	**Apr. 10**
THEATRE:	Silk Street
ADDRESS:	**£35**
CONCERT:	*8.30 pm*

DATE:	_____
TIME:	_____
PRICE:	_____
SEAT:	_____
THEATRE:	_____
ADDRESS:	_____
CONCERT:	_____

16 Match the verbs and nouns.

1 watch   a a photograph
2 cook   b music
3 take   c basketball
4 listen to   d a book
5 play   e TV
6 paint   f dinner
7 read   g money
8 spend   h a picture

17 Complete with prepositions.

Did you have lunch ___*at*___ Maxim's
when you were ___*in*___ Paris?

1 They live _____ a house _____ the
country.
2 We arrived ___*in*___ the airport _____
quarter past eleven.
3 We often go _____ the cinema _____
Saturday evenings.
4 I walked _____ and _____ the street,
but I didn't find the bank.
5 My birthday is _____ February.
6 My sister's birthday is _____ 18th May.
7 He left Chicago _____ 12 and
arrived _____ Atlanta _____ 2.

18 Write the times in the correct column.

yesterday   tomorrow   today
next year   last month   this week
every weekend   last night
tomorrow morning   two years ago
this afternoon   next Tuesday

Past	**Present**	**Future**
yesterday	_____	_____
_____	_____	_____
_____	_____	_____
_____	_____	_____
_____	_____	_____
_____	_____	_____
_____	_____	_____

19 Circle the odd-one-out.

run   (listen)   walk   swim
1 go   fly   wait   drive
2 cold   foggy   fine   black
3 ticket   date   bill   receipt
4 platform   train   kitchen   station
5 mountain   village   town   city

20 Which definition is correct?

a A train barks. ☐
b A person barks. ☐
c A dog barks. ☑
1 To climb is to
a go in. ☐
b go up. ☐
c go out. ☐
2 Terrible means
a awful. ☐
b great. ☐
c nice. ☐
3 Cricket is a
a computer game. ☐
b sport. ☐
c musical instrument. ☐
4 a You pay a bill. ☐
b You spend a bill. ☐
c You buy a bill. ☐
5 She got dressed means
a she wore a dress. ☐
b she bought a dress. ☐
c she put on her clothes. ☐

	Last week
Andrea	see a film
	have a party
Sebastian	go to a concert
	buy a new computer game
Mr and Mrs Hill	go to Amsterdam for the weekend
	travel to Amsterdam by coach

WRITING

21 Write about these people's activities.

Last week Andrea _____

_____

_____

Last week Sebastian _____

_____

_____

Last week Mr and Mrs Hill _____

_____

_____

Last week I _____

_____

_____

_____

22 Write about a weekend or day out you enjoyed. Try to answer some of these questions.

Where did you go?

When did you go?

How did you get there?

Who did you go with?

What did you do?

What did you see?

What did you like?

How long did you stay?

_____

_____

_____

_____

_____

_____

_____

_____

_____

_____

_____

_____

Answer key

Lessons 1-5

GRAMMAR

1   1 are, am 2 is 3 is 4 am 5 is 6 are 7 is, is 8 Is 9 is

2   1 You're Russian.

2 She's a student.

3 Steve's a doctor.

4 I'm from Turkey.

5 Tina's an actress.

6 What's your name?

7 He's Brazilian.

8 Monica's American and she's a singer.

9 Paulo's a waiter and he's from Rome.

10 What's your job?

3   1 a 2 b 3 c 1

2 a 3 b 4 c 1 d 5 e 2

4   1 What's your name?

2 How are you? I'm fine, thanks.

3 John's from New York.

4 He's from Tokyo and he's Japanese.

5 You're an engineer.

6 Suzanne's a teacher. She's from France.

7 Mr Ford's from London.

8 What's your telephone number?

5   1 b 2 a 3 a 4 b 5 b 6 a

6   1 a 2 an 3 an 4 a 5 an 6 a 7 a 8 a

7   What's your name?

How do you spell Bruno?

Are you an engineer?

What's your telephone number?

8   1 She's 2 She's 3 He's 4 She's 5 He's 6 He's

9   1 a 2 b 1 c 3

2 a 3 b 1 c 4 d 2

10 1 d 2 f 3 a 4 e 5 c 6 b

SOUNDS

12 1 Are you a <u>st</u>udent?

2 I'm from <u>Th</u>ailand.

3 What's your <u>name</u>?

4 What's your <u>tele</u>phone <u>num</u>ber?

5 She's a <u>doc</u>tor.

6 I'm very <u>well</u>, <u>thanks</u>.

7 <u>Thank</u> you very <u>much</u>.

8 He's from Ja<u>pan</u>.

9   I'm A<u>meri</u>can and I'm an eng<u>ineer</u>.

10 She's from New <u>York</u>.

13 □□ listen, waiter, student, zero

□□ correct, complete, hello, Brazil, Japan

□□□ underline, engineer, Japanese

□□□ punctuate, journalist, Italy, telephone

14 1 V 2 J 3 C 4 U 5 R

VOCABULARY

15 1 c 2 d 3 f 4 a 5 e 6 b

16 1 waiter 2 teacher 3 secretary 4 doctor 5 journalist

Hidden job: actor

17 1 Great Britain 2 Turkey 3 Brazil 4 Thailand 5 Japan

18 1 Russian 2 very well 3 Brazilian 4 job 5 nationality 6 Britain

7 country 8 he

20 1 a 2 b 3 c 4 b 5 c

WRITING

21 Anna is British. She's from Manchester and she's a journalist.

Marcos is Brazilian. He's from Rio de Janeiro and he's a teacher.

Murat is Turkish. He's from Ankara and he's a student.

22 *Example answer*

Walter	Hello, what's your name?
Anne	I'm Anne. What's your name?
Walter	I'm Walter. What's your job?
Anne	I'm a singer. What's your job?
Walter	I'm an actor. Where are you from, Anne?
Anne	I'm from Britain.
Walter	I'm American. I'm from the USA.

Lessons 6-10

GRAMMAR

1   1 They 2 I 3 She 4 they 5 She 6 he 7 They 8 He 9 I

2

I'm	I'm not	Am I?
You're	You aren't	Are you?
He's	He isn't	Is he?
She's	She isn't	Is she?
It's	It isn't	Is it?
We're	We aren't	Are we?
They're	They aren't	Are they?

3 1 he isn't 2 they aren't 3 we are 4 it is 5 she isn't
   6 they aren't 7 she is 8 it isn't 9 they are 10 you are

4 1 His 2 is 3 She's 4 Her 5 are 6 They're

5 1 What's this? It's a sandwich

   2 What's that? It's a telephone.

   3 What are these? They're watches.

   4 What are those? They're keys.

   5 What are these? They're pizzas.

   6 What's that? It's a bag.

   7 What's this? It's a wallet.

   8 What are those? They're books.

6 1 e 2 f 3 a 4 c 5 b 6 g 7 d

7 1 We aren't twins.

   2 It isn't an umbrella.

   3 Daniel isn't from Canada.

   4 Albert and Sandy aren't students.

   5 You aren't 25.

   6 His favourite singer isn't Elton John.

   7 They aren't Japanese.

   8 I'm not a politician.

8 1 My, My 2 your, my 3 your 4 your 5 your, My

9 1 his, His 2 her 3 her, Her 4 her, Her 5 his 6 his

10 1 Is she a teacher? Yes, she is.

   2 Are Tom and Gordon doctors? Yes, they are.

   3 Are you from Istanbul? No, I'm not.

   4 How old are you?

   5 What's your favourite car?

   6 Who's your favourite actor?

   7 The women are from Paris.

   8 What's this? It's a watch.

   9 Is Pauline a secretary? No, she isn't.

   10 She's Spanish. Her name's Maria.

11 1 Who 2 What 3 What 4 What 5 How 6 Where 7 How
   8 Where

SOUNDS

12 1 She's twenty-seven.

   2 Jane's American and she's an actress.

   3 Is he a student?

   4 Her favourite car is a Porsche.

   5 What's your address?

   6 My English teacher's from Los Angeles.

   7 I'm twenty and I'm from England.

   8 Is Marc French?

13 ☐   books, names, boys, fine

   ☐☐   glasses, pizza, watches, women, fifty

   ☐☐   cassette, repeat, address, translate

14 /s/: clocks, cassettes, wallets, books, students, groups

   /z/: glasses, twins, friends, umbrellas, televisions, bags, girls,
        brothers

VOCABULARY

15 1 eighty 2 fifteen 3 twenty-three 4 forty-five 5 thirty-three
   6 fifty-one 7 twenty-four 8 sixty-four 9 seventy-three
   10 twelve

16 1 books 2 wallet 3 clock 4 pen 5 telephone 6 glasses
   7 watch 8 bag

17 1 country 2 TV programme 3 umbrella 4 married 5 chair

18 1 cassettes 2 watches 3 men 4 twins 5 actresses 6 buses
   7 brothers 8 women 9 videos 10 football teams

19 1 from 2 man 3 seventy 4 How 5 group 6 French 7 married
   8 actress 9 women

   Hidden word: favourite

20 1 f 2 d 3 e 4 c 5 a 6 b

Lessons 11-15
GRAMMAR

1 1 Have, got 2 's got/has got 3 haven't got, 've got 4 've got
   5 Has, got, has 6 haven't got, 've got 7 Has, got, hasn't
   8 Has, got, has

2 | I've got | I haven't got |
   |---|---|
   | You've got | You haven't got |
   | He's got | He hasn't got |
   | She's got | She hasn't got |
   | It's got | It hasn't got |
   | We've got | We haven't got |
   | They've got | They haven't got |
   | Have I got? | Yes, I have. |
   | Have you got? | No, you haven't. |
   | Has he got? | Yes, he has. |
   | Has she got? | No, she hasn't. |
   | Has it got? | Yes, it has. |
   | Have we got? | No, we haven't. |
   | Have they got? | Yes, they have. |

3 1 my 2 her 3 their 4 your 5 our 6 his

4 1 d 2 f 3 e 4 a 5 g 6 b 7 h 8 c

5 1 Don't open your book.

   2 Don't stand up.

   3 Don't sit down.

   4 Don't put your pen down.

   5 Don't pick your bag up.

6 Don't put your jacket on.

7 Don't come in.

8 Don't close the door.

6　1 Has he got a watch? Yes, he has.

2 Has he got long hair? No, he hasn't.

3 Has he got a bag? No, he hasn't.

4 Has he got glasses? Yes, he has.

5 Has he got fair hair? Yes, he has.

6 Has he got a wallet in his hand? No, he hasn't.

7　1 a, a 2 The, the 3 – 4 a 5 the, the 6 an 7 – 8 a 9 –, –
10 the, the

8　1 Tim's wallet 2 Teresa's brother 3 Marina's children
4 Paul and Liam's books 5 Frank's sweaters 6 Anna's skirt
7 Ben and Eva's parents 8 David's sister 9 Manuel's father
10 Laura's teacher

9　1 is, 's got, is

2 's got, is, is

3 are, 've got, are

4 am, 've got, am

5 is, is, 's got

6 are, is, is

10　1 How old is he?

2 How are you?

3 What's she like?

4 How much is the white jacket?

5 Have you got any brothers (or sisters)?

6 Where is Nancy's bag?

7 What are they like?

8 Has he got a daughter?

11　1 They're Jane's keys.

2 The black shoes are on the chair.

3 How much are these jeans?

4 The green sweaters are £50.

5 George's bag is under the table.

6 They've got two brothers.

SOUNDS

12　1 <u>How</u> <u>much</u> are the <u>red</u> <u>shirts</u>?

2 They're <u>twenty</u> <u>pounds</u>.

3 <u>Dan's</u> <u>wall</u>et's <u>un</u>der the <u>table</u>.

4 <u>Kate's</u> <u>glass</u>es are in her <u>pocket</u>.

5 We've got <u>two</u> <u>sisters</u> and a <u>brother</u>.

6 Have you <u>got</u> any <u>children</u>?

7 He's got <u>dark</u> <u>hair</u> and <u>green</u> <u>eyes</u>.

13　1 red 2 jacket 3 book 4 wallet 5 husband 6 jeans 7 skirt
8 tall

VOCABULARY

14　1 There is a sweater in the bag.

2 There is a watch on the bag.

3 There are some sunglasses on the bag.

4 There is a book under the bag.

5 There is a jacket under the bag.

15　1 How much is the sweater?

2 How much are the shoes?

3 How much are the jeans?

4 How much is the chair?

5 How much are the sunglasses?

16　Traffic lights: red, orange/amber/yellow, green

American flag: red, white, blue

Piano: black, white

17　My brother's a doctor. He's got short hair and green eyes.
He's quite good-looking and he's very friendly. He's married
and he's got three children: a son and two daughters.

18　*Clothes*　jeans, jackets, shoes, trainers, sweater, skirt

Colours　black, white, green, blue, brown

Prices　£9.99, £50, £45.50, 50p, £6.80, £59.60

19　1 husband 2 mother 3 son 4 sister 5 father 6 daughter
7 children 8 wife 9 parents

20　1 pen 2 jeans 3 black 4 light 5 come 6 very

Lessons 16-20
GRAMMAR

1　1 Are 2 Do 3 have 4 are 5 Do 6 have 7 Do 8 are 9 Have
10 Do

2

I work	I don't work
You write	You don't write
We have	We don't have
They live	They don't live
Do I work?	Yes, I do.
Do you write?	No, I don't.
Do we have?	Yes, we do.
Do they live?	No, they don't.

3　1 lives 2 works 3 leaves 4 goes 5 has 6 starts 7 finishes
8 arrives

4　1 works 2 is 3 has 4 leaves 5 arrives 6 starts 7 has
8 finishes 9 goes 10 visits 11 corrects 12 watches

5　1 possessive 2 is 3 has 4 is 5 possessive, is 6 is, possessive
7 has 8 is, possessive 9 is, is 10 is, possessive

6　1 They don't like football.

2 Mike hasn't got a brother.

3 Karen isn't married.

4 We aren't Turkish.

5 They haven't got dark hair.

6 I don't work in Boston.

7 You don't have dinner at 9 pm.

8 My name isn't Bill Bailey.

7 1 to, in 2 in, in 3 at, at 4 on 5 at, in 6 to, in 7 to, on

8 1 I do 2 we don't 3 they do 4 she has 5 they aren't 6 I do 7 I haven't

9 1 Do you like sailing?

2 What is the time?

3 Are you married?

4 Do you have breakfast at 5 o'clock?

5 Does he go shopping on Fridays?

6 Where's she from?

7 Have you got a flat?

10 1 have 2 have got 3 has got 4 have 5 Have you got

11 1 We go shopping in the morning.

2 Her husband works in New York.

3 I go home at 7 pm.

4 I like basketball very much.

5 They have lunch at 1 pm.

6 We don't work on Saturday and Sunday.

SOUNDS

12 1 I <u>work</u> in an <u>office</u> in Ma<u>dr</u>id.

2 His <u>ch</u>ildren go to <u>school</u> in the <u>morning</u>.

3 Do you like <u>basketball</u>?

4 I like gymnastics <u>ve</u>ry <u>much</u>.

5 Do you <u>work</u> in a <u>school</u>?

6 What's the <u>time</u>, <u>please</u>?

7 They <u>live</u> in a <u>house</u> in Chi<u>ca</u>go.

13 Columns 1 and 2 have probably got the most words.

VOCABULARY

14 *on*   Sunday morning, Wednesday, Thursday evening, Sundays

in   the evening, the afternoon, the morning

at   three o'clock, a quarter past two

15 1 c 2 e 3 b 4 f 5 a 6 d

16 football, table tennis, baseball, basketball, volleyball

17 Tuesday, Wednesday, Thursday, Friday, Saturday, Sunday

18 1 e 2 h 3 d 4 a 5 c 6 b 7 f 8 g

19 a 5 b 2 c 7 d 4 e 1 f 8 g 6 h 3

WRITING

21 Wayne is American. He's a singer. He's got two sisters. He lives in a flat. He likes basketball and skiing.

Fiona is Scottish. She's a shop assistant. She's married, but she hasn't got any children. She lives in a house. She likes reading.

Mick and May are Australian. They're students. They're twins. They live at home with their parents. They like sailing.

Lessons 21-25

GRAMMAR

1 1 Does 2 don't 3 do 4 are 5 doesn't 6 Is 7 Have 8 are 9 isn't 10 does

2

I eat	I don't eat
You drink	You don't drink
He likes	He doesn't like
She lives	She doesn't live
It has	It doesn't have
We walk	We don't walk
They go	They don't go
Do I eat?	Yes, I do.
Do you drink?	No, you don't.
Does he like?	Yes, he does.
Does she live?	No, she doesn't.
Does it have?	Yes, it does.
Do we walk?	No, we don't.
Do they go?	Yes, they do.

3 1 she does 2 they don't 3 there isn't 4 she has 5 there are 6 he does 7 I don't 8 they aren't 9 it does 10 I do

4 1 Mary and Joe like dancing.

2 Patrick and Joe like reading.

3 Mary doesn't like reading.

4 Patrick doesn't like having parties.

5 Mary and Joe like having parties.

6 Patrick likes doing homework.

7 Mary and Joe don't like doing homework.

5 1 My parents often go sightseeing.

2 We always have English on Mondays.

3 What do you usually do in August?

4 Paul sometimes invites friends home.

5 I never write letters.

6 1 c 2 a 3 d 4 f 5 b 6 e

7 1 How does Neil go to work?/Who goes to work by train?

2 What time/When does Alexandra start work?

3 Where do they live?

4 What is there on the table?

5 How much is it?

6 How old is she?

7 What do you like (doing)?

9 1 b 2 d 3 e 4 a 5 c 6 f

10 *Example answers*

1 I live in …

2 I go to school/work by …

3 I have lunch at …

4 I usually have … for lunch.

5 I spend my birthday with …

6 Yes, I do/No, I don't.

7 I often …

11 1 a 2 c 3 b 4 b 5 a 6 c 7 a

SOUNDS

12 1 <u>Max</u> goes to <u>school</u> by <u>car</u>.

2 <u>When's</u> your <u>birth</u>day?

3 I <u>of</u>ten <u>go</u> to <u>res</u>taurants.

4 <u>What</u> do you <u>do</u> with your <u>friends</u>?

5 There's a <u>cup</u>board in the <u>hall</u>.

6 Are there any <u>peo</u>ple in the <u>class</u>room?

7 I <u>like</u> <u>stay</u>ing in ho<u>tels</u>.

13 1 cooker 2 fourth 3 go 4 front 5 do 6 out 7 on 8 word

VOCABULARY

14 1 yoghurt 2 Friday 3 twelve 4 beer 5 eating

15 *Example answers*

1 coffee, tea, orange juice

2 lamb, beef, chicken

3 wine, water, juice, milk

4 orange, apple, lemon

5 potato, tomato, French fries

6 bread, cereal, yoghurt

16 In my house there's a dining room, a large living room, a kitchen and a hall downstairs. Upstairs, there are three bedrooms and a bathroom. My room is at the back and there is a big window onto the garden.

17 1 14th February

2 4th July

3 1st January

4 25th December

18 1 eating 2 drink 3 beach 4 have 5 hotels 6 July 7 likes 8 dancing 9 sun 10 skiing

Hidden word: travelling

19 train, car, bus, taxi, boat, bicycle, plane …

20 1 a, c 2 a, b 3 b 4 a, b 5 c

Lessons 26-30
GRAMMAR

1 1 Kathy can't cook.

2 John can't speak Spanish.

3 Kathy can speak Spanish.

4 John can drive.

5 Kathy can drive.

6 John can play the guitar.

7 Kathy can't play the guitar.

8 John can't draw well.

9 Kathy can draw well.

2 Interviewer   Are you a secretary?

Tina   No, I'm not. I'm a student.

Interviewer   So you want a holiday job in an office?

Tina   Yes, I do.

Interviewer   Well, can you type?

Tina   Yes, I can. And I can use a computer.

Interviewer   Can you speak French?

Tina   No, I can't, but I can speak German.

Interviewer   Well, can you start tomorrow?

Tina   Yes, I can. Thank you!

3 I'm making   I'm not making

You're having   You aren't having

He's reading   He isn't reading

She's waiting   She isn't waiting

It's working   It isn't working

We're talking   We aren't talking

They're listening   They aren't listening

Am I making?   Yes, I am.

Are you having?   No, you aren't.

Is he reading?   Yes, he is.

Is she waiting?   No, she isn't.

Is it working?   Yes, it is.

Are we talking?   Yes, we are.

Are they listening?   Yes, they are.

4 sitting, running, drinking, lying, writing, having, standing, doing

5 1 I'd like a cup of coffee.

2 Anything to eat?

3 Thank you very much.

4 Can I help you?

5 Can I have a cheese sandwich, too?

6 Waitress   Can I help you?

Customer I'd like a cup of coffee, please.

Waitress   Anything to eat?

Customer Yes. Can I have a cheese sandwich, too?

Waitress Certainly. Here you are.

Customer Thank you very much.

7 1 I can 2 they aren't 3 they are 4 he can't 5 she is 6 I do
7 they haven't 8 there aren't 9 he does 10 they can

8 1 a 2 a 3 a 4 b 5 a 6 a 7 b

9 You use the present simple to talk about customs, habits and
routines. The present continuous is for something that is
happening at the moment.

10 1 Where's Robin from?

2 Where are they sitting?

3 What are you cooking?

4 Can you play the piano?

5 Where do you live?

6 What does she want for her birthday?

7 Does she like music?

11 1 I'm standing in a queue.

2 Are you buying lunch?

3 The men are sitting in a bar.

4 She doesn't want a sandwich at the moment.

5 My mother is shopping this morning.

6 Do you need help?

7 I can't speak Russian.

SOUNDS

12 1 I can <u>speak</u> <u>French</u>, but I <u>can't</u> <u>speak</u> <u>It</u>alian.

2 What about <u>sport</u>?

3 Can I <u>help</u> you?

4 <u>Anything</u> to <u>drink</u>?

5 A <u>piece</u> of <u>chocolate</u> <u>cake</u>, <u>please</u>.

6 <u>Where's</u> the <u>library</u>, <u>please</u>?

7 Turn <u>right</u> at the <u>bank</u> and go <u>straight</u> a<u>head</u>.

8 They're <u>getting</u> up and <u>having</u> <u>break</u>fast.

9 What are you <u>doing</u> at the <u>moment</u>?

13 1 swim 2 east 3 bank 4 lie 5 can

VOCABULARY

14 *Example answers*

1 Turn right into Market Street. Turn left into Park Street.
Turn right into King's Road and the post office is on your
right.

2 Go along Castle Street. Turn left into Churchill Avenue.
The car park is on your left.

3 Go straight ahead and turn right into Queen's Street. Go
past Park Street and the bookshop is on your left.

15 1 The pub 2 The station 3 The bank

16 *Sandwiches* tuna and mayonnaise, cheese and tomato,
chicken and lettuce, cheese

Hot meals pasta, beef with rice, pie with vegetables, pizza

Desserts yoghurt, chocolate cake, ice cream, apple pie

Drinks milk, coffee, cola, tea

17 1 piece 2 glass 3 bottle 4 cup

18 *Example answers*

1 have a shower/have breakfast with my family.

2 have toast and a cup of coffee.

3 have a sandwich and some yoghurt. I drink fruit juice.

4 meet my friends in a café.

5 watch TV/play with my children.

19 *Example answers*

I can't make apple pie.

I can use a computer.

I can't play the piano.

I can speak English.

20 1 You watch a film in a/the cinema.

2 You have a bath in the bathroom.

3 You buy books in a/the bookshop.

4 You park your car in a/the car park.

5 You take the train at the station.

WRITING

22 *Example answers*

Gordon is eating a sandwich and reading a book. Lucy is
talking on the phone. Julie is listening to music and drinking
a cup of coffee/tea. Ben is playing the guitar and singing.
Sam is cooking.

Lessons 31-35
GRAMMAR

1 1 c 2 a 3 d 4 e 5 b

2 Amanda Let's go to the theatre this evening.

Tony Oh no! I'm sorry, but I don't like the theatre.

Amanda Well, how about going to the cinema, then?

Tony What's on?

Amanda The new Tom Cruise film.

Tony And where's it on?

Amanda At the Odeon Cinema.

Tony OK, then.

3 1 On Tuesday, she's having lunch with Larry.

2 On Wednesday, she's going shopping with Sarah.

3 On Thursday, she's getting tickets from the travel agent's.

4 On Friday, she's visiting her parents.

5 On Saturday, she's flying to New York.

4   2, 5, 6

5   1 When is the/her plane leaving?

2 What is Maria's mother doing?

3 What are the people (in the Italian restaurant) doing?

4 Where are you going this evening?

5 Where are your brother and his wife going?

6
I was	I wasn't	Was I?	Yes, I was.
You were	You weren't	Were you?	No, you weren't.
She was	She wasn't	Was she?	Yes, she was.
We were	We weren't	Were we?	No, we weren't.
They were	They weren't	Were they?	Yes, they were.

7   1 was 2 weren't 3 Were 4 was 5 were 6 was 7 was, wasn't 8 Was 9 were

8   1 b 2 c 3 a 4 b 5 b 6 a

9
I had	I didn't have	Did I have?	Yes, I did.
You had	You didn't have	Did you have?	No, you didn't.
He had	He didn't have	Did he have?	Yes, he did.
We had	We didn't have	Did they have?	Yes, they did.

10   1 Did it have a television? Yes, it did.

2 Did it have a video recorder? No, it didn't.

3 Did it have a fax machine? No, it didn't.

4 Did it have a table? Yes, it did.

5 Did it have a telephone? Yes, it did.

6 Did it have a computer? No, it didn't.

11   1 Where are you going next week?

2 I was in Paris yesterday.

3 She doesn't have breakfast at home every morning.

4 We were tired last night.

5 My sister likes pop music.

6 Her teacher doesn't work on Fridays.

7 The boys had a party last Saturday.

8 Her parents are getting the plane next Sunday.

SOUNDS

12   1 <u>Was</u> she in the <u>kitchen</u>?

2 <u>No</u>, she <u>wasn't</u>.

3 <u>Let's go</u> to the <u>theatre</u>.

4 I'm <u>sorry</u>, but I <u>don't like films</u>.

5 <u>Where's</u> it <u>on</u>?

6 The <u>film</u> was <u>awful</u>.

7 The <u>radio programmes</u> were <u>good</u>.

13   radio, dishwasher, bicycle, grandparents, cameras, families, garden, thirsty, butler, happy, yesterday, theatre, anyone

VOCABULARY

14   1 packing 2 meeting 3 camping 4 wearing 5 getting 6 spending 7 flying 8 arriving

16   1 bored 2 hot 3 hungry 4 thirsty 5 Happy 6 cold 7 tired

18   1 c 2 a 3 e 4 b 5 d

19   a 4 b 7 c 1 d 2 e 9 f 6 g 5 h 3 i 8

20   *Example answers*

Verbs meet, go, see, wear, get, arrive …

Nouns   city, beach, gallery, film, office, kitchen …

Adjectives   hot, cold, tired, happy, bored, awful …

Prepositions   in, on, at, with, to, by …

Pronouns   I, you, he, she, we, they …

Lessons 36–40

GRAMMAR

1
looking	looked
having	had
buying	bought
getting	got
coming	came
studying	studied
taking	took
stopping	stopped
going	went
dancing	danced
leaving	left

2
I stayed	I didn't stay
You flew	You didn't fly
He ate	He didn't eat
She saw	She didn't see
We found	We didn't find
They read	They didn't read
Did I stay?	Yes, I did.
Did you fly?	No, you didn't.
Did he eat?	Yes, he did.
Did she see?	No, she didn't.
Did we find?	Yes, we did.
Did they read?	No, they didn't.

3   1 arrived 2 took 3 went 4 bought 5 had 6 visited 7 saw 8 spent 9 were 10 arrived

4   1 Where did they fly from?

2 When did they arrive?

3 How did they get from the airport?

4 What did they do on Saturday afternoon?

5 What did they buy?

6 Where did they go on Sunday?

7 How long did they spend there?

5 1 ago 2 this 3 last 4 every 5 next 6 today

6 1 I did 2 she didn't 3 I wasn't 4 it was 5 they didn't
6 she does 7 they have

7 1 irregular 2 frequency 3 quantity 4 modal 5 simple
6 imperatives 7 articles

8 1 Rick wasn't born in Milan.

2 Rick didn't go to college in Rome.

3 Rick doesn't like sport.

4 Rick doesn't drive a Fiat.

5 Rick hasn't got a big family.

9 1 did you spend 2 are playing 3 didn't sleep 4 was
5 Do you often go 6 lived 7 are you doing 8 likes

10 *Present simple*   (5) Do you go (8) likes

    Present continuous   (2) are playing (7) are you doing

    Past simple   (1) did you spend (3) didn't sleep
                            (4) was (6) lived

11 *Example answers*

1 I was born in…

2 I went to school in…

3 I played … at school.

4 I play … now.

5 The last time I went on holiday was…

6 I went to…

7 I stayed…

8 I'm going to … for my next holiday.

SOUNDS

12 1 He <u>went</u> to <u>Lon</u>don.

2 We <u>left</u> our ho<u>tel</u> at <u>ten</u> o'<u>clock</u>.

3 What did they <u>do</u> in the after<u>noon</u>?

4 <u>How</u> <u>much</u> were the <u>tick</u>ets?

5 She <u>found</u> the <u>cat</u> in her <u>gar</u>den.

6 We <u>went</u> <u>home</u> in the evening.

7 When he was fif<u>teen</u>, he <u>left</u> <u>Ma</u>laga.

13 /t/ watched, worked, cooked, danced, finished

/d/ stayed, travelled, played, listened, lived, died, called,
happened

/ɪd/ wanted, visited, started, painted

VOCABULARY

14 1 city centre 2 computer games 3 team sport 4 waiting room
5 ticket office 6 dining room 7 dishwasher

15

Date	Apr. 10
Time	8.30 pm
Price	£35
Seat	Circle D10
Theatre	Barbican Hall
Address	Silk Street
Concert	Jazz Scenes

16 1 e 2 f 3 a 4 b 5 c 6 h 7 d 8 g

17 1 in, in 2 at, at 3 to, on 4 up, down 5 in 6 on 7 at, in, at

18

Past	yesterday, last month, two years ago
Present	today, this week, every weekend, this afternoon
Future	tomorrow, next year, last night, tomorrow morning, next Tuesday

19 1 wait 2 black 3 date 4 kitchen 5 mountain

20 1 b 2 a 3 b 4 a 5 c

WRITING

21 Last week, Andrea saw a film and had a party.

Last week, Sebastian went to a concert and bought a new
computer game.

Mr and Mrs Hill went to Amsterdam for the weekend. They
travelled by coach.